Universe

6 books in 1

Andrew Williams

Copyright © 2016

Copyright © 2016 Andrew Williams

All rights reserved. This book or any portion thereof may not be reproduced or used in any manner whatsoever without the express written permission of the publisher except for the use of brief quotations in a book review.

Disclaimer

This book is designed to provide condensed information. It is not intended to reprint all the information that is otherwise available, but instead to complement, amplify and supplement other texts. You are urged to read all the available material, learn as much as possible and tailor the information to your individual needs.

Every effort has been made to make this book as complete and as accurate as possible. However, there may be mistakes, both typographical and in content. Therefore, this text should be used only as a general guide and not as the ultimate source of information. The purpose of this book is to educate.

The author or the publisher shall have neither liability nor responsibility to any person or entity regarding any loss or damage caused, or alleged to have been caused, directly or indirectly, by the information contained in this book.

Book List

CAN HUMAN BEINGS REACH THEIR FULL POTENTIAL? 4

DO PARALLEL UNIVERSES REALLY EXIST? 20

CAN HUMAN BEINGS LIVE AND SURVIVE ON DIFFERENT PLANETS? 32

IS TELEPORTATION POSSIBLE? 44

IS TIME TRAVEL POSSIBLE IN OUR UNIVERSE? 60

ARE UFO PILOTS REALLY ALIENS? 74

Can Human Beings Reach Their Full Potential?

Introduction

You can achieve your maximum potential. But, do you even know what you are capable of doing once you achieve this? What you might not know about is that the mind and body are equipped with a limiter. As the term obviously means, this limits the human body from exerting greater force than it is usually capable of.

For example, have you ever heard of a person suddenly being capable of carrying a huge appliance like a refrigerator? You might not believe it if the person is a scrawny looking dude who has never even visited the gym to work out. He is, in his full potential, actually capable of lifting an object that can be ten times heavier than his weight. He can possibly do this because of adrenaline. What happens when a person is in danger or is facing a potential dangerous situation is that the brain releases adrenaline to boost his body function. This, in effect, temporarily removes the limiter.

You would then ask why the limiter exists. Wouldn't it be better if everyone is allowed to have more "power?" One of the reason that limiter exists is to protect the human body. Too much exertion of force can destroy the body. A good example is a boxer. If a boxer does not use gloves or control his punches, every punch that he will land on his opponent can destroy his knuckles.

Aside from the human body's physical "powers," you can also think about maximizing your brain power to become a higher being such as becoming an esper, a psychic, a time traveler, or even a god. According to a popular myth, human beings are only using 10% to 11% of their brain, and the remaining 89% to 90% are untapped or unused. This is the part of the brain that has to be utilized to achieve the true potential of human beings.

Despite the controversies that debunk this myth, much attention is still devoted to the possibilities of wielding various powers that emanate from the brain and involve mental process. Studies continue to be done to prove that they exist. Classes and seminars are also held to help people tap into the unused parts of their brain. A lot of these classes take things further than mere learning. They promote the practice of the powers they discover and the transformation into a higher being.

In this book, you will get more information about these powers that are believed to be possible once you maximize your full potential as a human being. A few of these powers are the power to communicate with other unworldly beings, power to heal, power to defy laws of physics, and even the power to teleport.

Healing Powers

Energy medicine
Energy medicine, or less popularly known as spiritual healing, energy healing, and energy therapy, is the ability to use one's energy to heal another person. The three common ways used by a person with this power to apply energy medicine are: by touching the patient, standing by the patient, or being far from the patient. In contrast to faith healing, energy medicine does not need a patient or healer to follow certain religions for the beneficial effects to happen.

Out of the three ways to apply energy medicine or channel energy, the distant or absent energy healing is the most controversial. Due to its popularity and ineffectiveness, it has been the subject of government scrutiny. Unfortunately, most of the distant energy healing practitioners were considered fraudulent. Alternatively, the two other methods were able to show some positive results. Nevertheless, people are skeptical if it is truly the healing energy is healing the patients, or the patient is only experiencing a placebo effect.

Faith healing
Faith healing is almost the same as energy medicine. However, they differ in a sense that faith healing requires a strong belief in a certain religion, and the only means of true healing are by a ritual or praying. Also, instead of using energy to heal a patient, the healer invokes a divine power or a divine presence to heal his patient.

Faith healing does not necessarily mean that a person needs to visit a person that has the ability to heal. He can heal himself by visiting a religious establishment, praying intently, or strengthening his belief or faith to the deity, god, or Supreme Being he worships.

Throughout history, many instances of recovery caused by faith healing have been recorded. Most of them were considered miraculous especially since they involve the healing of diseases that are considered incurable. To add, some of the diseases and ailments that some faith healing practitioners claimed that they can heal are injuries, skin rashes, AIDS, corns, defective speech, deafness, blindness, anemia, cancer, developmental disorders, arthritis, multiple sclerosis, and total body paralysis.

Power to See

Clairaudience

Clairaudience is a subtype or a different method of clairvoyance. People with the power of clairaudience can hear messages or voices in a paranormal manner. For example, they are capable of hearing the voices of dead people. Individuals with clairaudience are typically referred to as people with third ear. And they are sometimes mistaken as mentally ill persons.

An individual who has this gift does not necessarily hear actual audible sounds. Sometimes, they can just "hear" these things in their thoughts. Technically, it is the same as talking to oneself without speaking. However, some can hear audible sounds that other people are not able to hear.

Clairvoyance

Compared to clairaudience, clairvoyance is more popular. Clairvoyance is the ability to gain information about people, objects, locations, and events through forms that the regular human senses cannot capture. Usually, this psychic power is often known by many as the ability to see the future.

Some parapsychologists suggested that if human beings discovered how clairvoyance works, it is possible for them to unlock other psychic abilities such as precognition and telepathy. As of now, many psychics claim that they are clairvoyant, but none of them are able to provide solid proof that they have the power to do so.

Power to Locate

Dowsing
Dowsing is a very popular method of finding objects and essential resources during medieval times. Usually, a person who is dowsing use a Y or L shaped twig or metal to detect the things he is looking for. This tool is commonly called a divining rod, a dowsing rod, a vining rod, or a witching rod. However, some dowsers do not use any equipment at all. According to practitioners, they make use of ley lines (currents of Earth's radius) to detect the things they are searching. As of today, the practice of this ability is still popular among those people who still believe in radiesthesia and Forteana.

Multiple tests have been given to modern dowsers for the past decades. However, most of them failed to satisfy the test makers. Apparently, many have been profiting from people who believe in this ability or practice's effectiveness by selling expensive dowsing devices.

Remote viewing
Those people who are capable of sensing and viewing hidden or faraway objects, people, events, or places are said to be capable of remote viewing. Multiple private and government researches were done to study this ability ever since it gained popularity in the 1970s. At first, research institutions were able to gain significant data from their test subjects. However, when they tried to make their investigation systematic under controlled environments, results became unfruitful. Sadly, no one was able to come up with good outcomes from their researches and tests about remote viewing.

Scrying
Fortunetellers or psychics who check out their crystal balls to see or predict something have the ability of scrying. Technically, scrying allows a person to see anything that he or she desires to see by using an object that is reflective (e.g., mirrors, water, crystals). In addition to being able to view anything from anywhere, persons who can scry are also able to see events from the past, present, or future.

Most of the recorded events that involved the practice of scrying are considered as folklores. One of them is the tale of Bloody Mary. Another famous tale that demonstrated this power is the story of Snow White.

Teleportation and Gravity Defiance

Apportation
This ability allows an individual to teleport things or objects from one place to another. Sometimes, apportation can be referred to as alchemy, summoning, and materialization. In the real world, this psychic ability is a very famous magic trick done by performers. The most common apportation technique is making a rabbit or a dove appear from a magician's hat. Unfortunately, many have been apprehended because of fraudulent activities using this psychic ability.

Bilocation
Bilocation is a certain phenomenon wherein a person can exist in two different locations at the same time. Sometimes, a person with the capability to perform or experience this can exist in more than two places – hence it is sometimes called as multilocation. In some cases, this ability is also considered as an advanced form of astral projection.

Most of the people who exhibited this phenomena or power are saints and benevolent individuals of the Christian church. A few of them are St. Anthony of Padua, St. Ambrose of Milan, St. Severus of Ravenna, St. Pio of Pietrelcina, and St. Alphonsus Liguori. During their state of bilocation, people were able to confirm their presence in both locations. And it seems that the common factor that they share is that they were praying or attending the mass while their 'other self' is present on a location far away from their current position.

Astral projection
This is one of the most known psychic abilities in this world, and it is commonly associated with mental projection. Fundamentally, astral projection allows a person to travel outside his physical body. Technically, the astral body, or spirit/soul as some call it, will be capable of moving around the astral plane. Astral projection or astral travel is related to OBE (Out of Body Experience). People who were able to astral project are often asleep, ill, under the influence of drugs, in sleep paralysis, or having a near death experience.

According to a survey made during the 1960s to 1970s, almost 8% up to 50% of people included in the survey were able to experience astral projection. Although, for experts like Robert Tod Carroll and Bob Bruce, members of the Queensland Skeptics Association, astral projection is just a dream state or an illusion made by imagination.

Transvection

Transvection is the ability to fly or levitate. As of late, many street magicians have attempted to show this power to their audiences. However, most of them only received skeptical feedback. Nevertheless, no strong case or evidence has been recorded that can at least prove that transvection or human levitation is possible.

Power to Predict

Death-warning
Death warning is the capability to receive warnings about a dying person. Commonly, the warning that people with this ability receive is an apparition of the dying person. Alternatively, some receive warning in the form of sensory hallucination. Sometimes, they might just experience certain massive sensations or extreme depression.

Most of the people who experience receiving death warnings are healthy people. According to a census performed in 1889, almost 2.3% of the individuals who responded in the census were able to receive an accurate death warning.

However, the Society for Psychical Research stated that this ability is still questionable. The members of the said Society deemed that evidences that are more feasible are required to consider death warning as a legitimate ability that humans can possess. Since most of the data gathered for surveys regarding death warning are usually in a form of testimonies, one cannot ascertain whether the data is accurate or inaccurate.

Precognition
Precognition is also known as the ability to have precognitive dreams or to obtain premonitions. Parapsychologically speaking, it is the power to see the future or the ability to have future sight or second sight. Scientific experts do not believe in this ability strongly. Well, that is because of causality. Logically, the effect should not come first before the cause. Unless, the future is already written, there is no way that people will be able to see the future.

Nevertheless, a few case studies have obtained a good amount of data that can prove that precognition exists. In selected observations, researchers were able to determine that 34% – 75% of their subjects were able to experience precognitive dreams. However, many people doubt the existence of precognition.

Retrocognition

In contrast to precognition, retrocognition allows a person to gain insight from the past. Of course, the only way retrocognition is useful is to discover unknown past facts. And for that same reason, identifying the legitimacy of retrocognition becomes too difficult. Even if that is the case, many people support the existence of retrocognition.

One of the most believable cases of retrocognition is the vision of Eleanor Jourdain and Annie Moberly, who are both respectable scholars in a woman's university. In 1911, both have experienced an event, which seemed like a brief time travel to the past. During that time, they were able to see firsthand the last days of Marie Antoinette, Queen of France and Navarre. According to historians, their depiction of the queen and the historical place where the queen was were very accurate.

Psychometry

Psychometry or psychoscopy is a psychic power that is relatively similar to clairvoyance and clairaudience. The main difference is that a person with this ability needs to touch an object to gain information. Once a psychometric touches an object, he will be able to know or gain information related about the object. For example, if a psychometric touches a doll, he will know who made the doll, how it was shipped, who its owner is, and the events the doll was involved in.

Power of Advanced Communication

Aura reading
This power can make an individual detect or see a person or entity's aura. An aura is depicted as a faint glow that radiates from living and even nonliving objects. The aura can provide the aura reader about an individual's power level, energy level, emotional state, and morality. Ordinary human beings or people without ESP cannot discern auras using their naked eyes.

As usual, controversies surround people who practice aura reading. People tested if aura readers do have paranormal powers during many instances. One of the tests they used to discern if aura readers were not fakes is to let them identify humans from mannequins. Sadly, most of those aura readers failed to demonstrate their powers. On the other hand, skeptics concluded that those people have only accidentally saw auras due to the aid of psychedelic drugs such as LSD and cannabis. Alternatively, they might have cognitive and vision disorders like migraines, epilepsy, synesthesia, eye fatigue, and eye burn.

Telepathy
Telepathy is the ability to transfer thoughts and/or emotions to another individual without the need of physical or verbal interaction. There are four types of telepathy: latent telepathy, emotive telepathy, superconcious telepathy, and intuitive, retrocognitive, and precognitive telepathy. Multiple events were able to demonstrate that telepathy is entirely plausible and possible.

Power to Use Spirits and Extraterrestrial Beings

Automatic writing

Automatic writing is the power that permits a person to write something unconsciously or by allowing his subconscious, spiritual sources, or external beings to take over. Many automatic writers have been noted in history, and most of them claimed that they acted as mediums to spirits and external beings.

Experts claim that it is possible that this power only displays an ideomotor effect related to writing. The ideomotor effect, by the way, is a psychological phenomenon that makes a person perform something out of their will. A perfect example of the ideomotor effect is the tear that may come out from someone's eye even though he does not want to cry.

On the other hand, it is possible that those people who perform automatic writing are experiencing cryptomnesia. Cryptomnesia is a state wherein a person remembers a memory, but fails to recognize it as a memory and instead considers it as a new idea or experience. A good example is the case of Helene Smith, which was investigated by Theodore Flournoy, a psychology professor. Helene Smith claimed that she is an automatic writing medium. And Theodore Flournoy checked out the articles she wrote and claimed that they were messages from an entity from Mars. Apparently, Flournoy concluded that the thing she wrote might have been derived from her past memories that she was not able to identify.

Divination

Divination lets an individual gain insight about events, people, or questions by performing occult rituals. If one is gifted or capable, he will be able to receive the insights he wants. But if not, nothing will happen. Technically, different cultures and regions have dissimilar ways and interpretations of divination. To an extent, Julian Jaynes has created and separated divinations into four types; they are Spontaneous, Augury, Sortilege, and Omens.

Mediumship or channeling

Persons who can communicate with spirits and other dead human beings are called mediums, and they practice the ability called channeling or mediumship.

To communicate with the dead, mediums often concentrate first and try to call the dead person's name. In the event that he is able to get the attention of the person they are trying to contact, he will allow the spirit to possess him or try to listen and relay his messages.

Since ancient times, humans have made many attempts to call or talk to the dead. Again, just like any the other psychic abilities listed here, it gained a negative credibility due to fraudulent people. Up until now, many con artists are still on the loose and are fooling people about their ability to communicate with the dead. However, even if that is the case, many people still believe and hope that some espers are capable of being mediums.

Destructive Power

Psychokinesis

Psychokinesis is the power to move objects or people using mental power. This popular psychic ability is very prominent in science fiction stories involving the paranormal. In real life, only a few cases of people capable of demonstrating this power have been recorded. In those events, little distortions or movements of the objects being controlled were noticed. And that is the reason why the scientific community is still skeptical in believing its legitimacy. Another issue of those demonstrations is that most of the psychics who performed psychokinesis were unable to repeat the demonstration, and lacked the control over the said ability.

Pyrokinesis

With pyrokinesis, a person can control or manipulate fire. Even if pyrokinesis was derived from a work of fiction made by Stephen King, a few people were able to show that it is possible. Around in March 2011, a certain 3 year old girl in the Philippines was able to ignite a pillow by just saying the words 'fire pillow.' Local firemen and policemen were able to witness the girl causing or predicting fires. Aside from that case, most of other pyrokinesis demonstrations were shot down as frauds.

Conclusion

So there you have it. A normal person will naturally also be skeptical about these powers. Technically, they are very sketchy, and the current laws in the realm of science do not agree that that these powers can possibly exist.

Nevertheless, it is safe to say that the possibility is there. The only thing clear is that the scientific community does not agree with these concepts of current laws and boundaries of the human world. In the end, they were not able to prove or to disprove anything.

Do Parallel Universes Really Exist?

The Question Comes

When was the first time you thought about the possibility of parallel universes? Was it after reading a book or watching a movie? It is highly likely that your answer will be a yes. This idea is often encountered in the world of science fiction. In this realm, there will never be a shortage of works that tackle the topic of a multiverse. These stories feed the mind with questions about whether or not another reality exists.

If you've only stumbled upon a single book or film that involves parallel universes, then you might think that literary masters and movie makers all share the same concept. You need to keep in mind that the multiverse, which is also called meta-universe, is presented in different ways. To make that point much clearer, it is necessary to provide a few examples.

In *The One* (a movie that stars Jet Li), the protagonist and antagonist are essentially one and the same. They only differ in one key aspect – their homes. To be a bit more specific, the hero and the villain came from their own respective universes. So, what makes the film's interpretation of the meta-universe interesting?

It highlights the possibility of killing your own other-universe self to achieve greater power. If that seems too complicated and farfetched, then you might be more interested in *Uncertainty*. In this movie, viewers are merely encouraged to believe, although not directly, that a single coin flip could actually create two very different parallel universes.

If you're into books, then you might have heard of *Glory Road*. In that literary work, one particular invention makes trips throughout the multiverse possible – the Burroughs Drive. Instead of presenting the meta-universe concept as something that involves minor changes though, the book pitches a fascinating idea: the multiverse itself could just be a manifestation of fiction.

There is one other piece of literature that offers a much more scientific take on parallel universes. The book *The God Themselves* points out that one universe could differ from another in terms of their physical laws. In the book, humanity didn't simply find a way to travel between universes. The human race discovered a method of harnessing energy from physical-law variances.

Parallel universes are also an integral part of comic books. In particular, both Marvel and DC titles feature storylines that span across universes. Here's what makes that a lot more intriguing: the greatest minds of the comic-book industry were the ones who came up with the term "Omniverse." This is a word that refers to the existence of multiple multiverses.

If comic books are basically simpler novels, then television shows are the less extravagant siblings of full-length films. The humble TV offers many episodic tales about parallel universes as well. *Sliders* (which is about a lost boy who's traveling through the multiverse for the sake of finding his home) and *Fringe* (a complicated story that's told on two parallel universes) are merely two examples.

At this point, at least one thing should be clear – there's no denying that the meta-universe concept exists within people's minds. Why this is so should be the next question to be answered. What are the most likely reasons why humanity often comes up with thoughts on what could be seen on the other side (or on the many other sides) of reality?

Why People Wonder

Why do people wonder about the multiverse? Why do you think of things that might be happening (or might have happened) in another world that's just like yours? There are many answers to these questions. There are many reasons why every person comes up with thoughts about parallel universes. Some are simply interested in alternate history, while others imagine a reality where they made all the right decisions. Likewise, there are those who use the meta-universe concept either to criticize or to praise.

The Historical Perspective

Humans simply like to speculate about history. Asking "what if a historical event didn't happen?" is a natural tendency. It might sound like a digression from the topic of parallel universes, but historical speculation is often associated with time travel. From a logical standpoint, a trip to the past is necessary to change the present. You need to keep in mind though, that a different flow of historical events, no matter how slight, might have unfolded in a parallel universe.

In another world similar to this one for example, Hitler might not have risen to power. He might have been accepted as a student in the Vienna Academy of Art and would thus be able to conquer the world with his artistic talents and not with military power. Or, did you know that most of the tyrant's siblings didn't have the chance to enter adulthood? What if in one reality, he was among those who died early?

You've probably thought about those questions right after you learned about Hitler's infamy. That's how natural it is for people to ponder upon the multiverse. Some even think of parallel universes way before they learn about the term. It's true though, that asking "what if?" doesn't always involve historical figures. That question could also be about something personal.

Something Personal

Much like those around you, regrets are most likely a part of your life. Regardless of how few or minor they might be, your regrets have effects on your life. It's only normal for you to imagine what could have been, or more specifically what might be happening in another reality – a universe in which you've made another choice. In a way, you sometimes rely on the multiverse notion when you need to reflect on your own mistakes.

Supporting Advocacies

The idea of a parallel universe is also used to highlight an alternate scenario that supports a particular advocacy. One way through which critics of drug companies highlighted one of humanity's biggest errors is to illustrate a world where Thalidomide wasn't hurriedly released into the market. In this parallel world, a lot of people did not end up suffering from all sorts of deformities.

If you are not familiar with Thalidomide, it's a medication currently used to treat those with myeloma or leprosy. Decades ago, it was marketed as both a sedative and a morning-sickness fighter. While it did its job quite well, it came with a very serious side effect – pregnant women who took it gave birth to babies with physical abnormalities; some children even had flippers instead of arms.

On the other side of the fence, the idea of a multiverse could also be used to point out humankind's greatest feats. Advances in environment protection are lauded by showing a parallel universe with everything that these developments have prevented. In a parallel universe where people did not believe in the importance of shifting towards sustainability, natural-resource problems and environmental concerns would be unsolvable.

Scientific Perspective

If all people think about the meta-universe at one point or another, then it's safe to say that even the brightest minds in science spend some time pondering upon this fascinating topic too. Instead of merely focusing on historical and personal what-ifs though, scientists, or physicists in particular, are more methodical about answering the question of the existence of a parallel universe.

From the scientific perspective, there are four different intriguing answers to this question:

Multiple universes do exist, mainly due to the ever-expanding nature of the cosmos.
In other words, ever since the Big Bang occurred, the number of new universes never stopped growing. You'd be surprised to learn though, that scientists who support this theory do not believe that every universe formed is very similar to ours. They hypothesize that each world has its own unique laws of physics. That means that in some parts of the multiverse, life couldn't exist. On the other hand, there are regions in the greater cosmos that almost have the same physical laws as this world.

If those parallel universes do exist, then they could be seen, right? Well, according to the theory's proponents, it would be impossible to see those worlds because most of the other universes are simply too far away. The ones that are supposedly near enough to be seen are inside black holes. In a way, it's a bit challenging not to think of those explanations as mere excuses.

This particular reality (or universe) is merely a three-dimensional surface within a nine-dimensional space.
Simply put, there's so much space available throughout the greater cosmos that there would probably be other three-dimensional surfaces. Those two parallel-universe ideas aren't really well liked by scientists who delve into quantum mechanics. Such scientific experts prefer to believe in the Many-Worlds Theory that the multiverse concepts in many works of fiction are patterned after. A new question emerges from this: what makes that idea much more

interesting than its counterparts? The answer to that lies in one word – choices.

The Many-Worlds Theory points out that choices create alternate timelines, which are real in every possible way. In that sense, if you once had an opportunity to choose between two universities and you picked university A, then in another universe you would actually end up going to university B. It should also be mentioned that even the most simple of choices could create new realities.

Much like in other theories, there's no way to see what's happening in another part of the many-worlds multiverse. You'll only be able to observe and experience events that unfold in your own timeline. Your other selves or those in other universes would have their own set of events and experiences. The similarities among the different events and experiences would depend on how distantly branched the timelines are from one another.

It's obvious that even the brightest minds in science could only speculate about parallel universes. When it comes to studying the subject of the multiverse, they only come up with theoretical explanations that are, at the very least, somewhat believable. While there's nothing wrong about focusing on the hypothetical, that approach has all sorts of flaws and downsides.

About the Opposition

You would probably notice that the list of theories in the previous chapter is incomplete; you were, after all, promised four intriguing theories. Well, here are the four theories.

Parallel universes don't belong in scientific endeavors.
For obvious reasons, this theory was left out of the list of theories about the existence of parallel universes. Scientists who are against the main meta-universe theories often emphasize the sheer impossibility of testing those grand notions.

As you've discovered a few pages back, there's no way to see what's unfolding in another universe. If you've been reading all kinds of scientific studies, then you'd know that the lack of capabilities to test these theories is a big problem. Testing, everyone knows, is the foundation of scientific research.

The Question of Evidence
When it comes to the field of science, evidence is the most important factor in either proving or disproving a theory or a particular set of assumptions. In order to get enough evidence, scientists need to carry out studies that would let them gather actual data. What makes that info-collection process really challenging though, is the need for repetition.

It would not be enough for one team of researchers to successfully gain supporting evidence. Discoveries and conclusions need to be checked against much newer data, which in turn means that countless studies regarding the same topic need to be done. So, even if someone claims that there's finally evidence for the multiverse's existence, that proof will be scrutinized.

You have to keep in mind that the problem of repetition or thorough confirmation isn't the only reason many scientists remain skeptical about the parallel-universe concept. They are also worried about the eagerness of their fellow experts when it comes to believing in something that could only exist in theory. Skeptics tend to assume that the technology to see other universes would never come.

Considering Chance

While it should be fine for theorists to mainly focus on the theoretical, embracing the multiverse notion isn't a matter of choosing the most logical hypothesis. In particular, many of those against the meta-universe idea usually argue that its proponents are willingly forgetting about the significance of chance in scientific endeavors.

Some would ask, "Isn't chance an integral part of the parallel-universe concept?" The formation of new physical laws and the need for options to emerge are considered as random factors. However, they would only be valid if the much grander theory of the continuous expansion of the grand cosmos holds true.

In addition to that, theorists have also made all sorts of assumptions for the sake of fine-tuning the multiverse and making the entire idea completely reasonable. As it turns out, the need to make things much more scientific and logical had an opposite effect. Believing in parallel universes, to an extent, has become similar to having faith in intelligent design.

This means that the universe didn't emerge out of sheer chance or randomness. The cosmos was made by a powerful being, or for most people by a God. In the next chapter, you will find out more about the link between two frequently -discussed topics – creationism and the multiverse.

The Issue of Creation

The question of whether or not creationism supports the parallel universe idea has been raised many times. There are similarities between the two that seem to indicate that it does. The need for fine-tuning for things to exist and the lack of evidence-gathering opportunities are common between the two. While it would be easy enough to jump into the conclusion that the two ideas support each other, the real answer to this question is actually no. Creationism does not support the parallel universe idea.

Scientists who support the meta-universe idea would never admit that their own fine-tuning attempts are in any away related to intelligent design. After all, if they do say that there's a link between the two, the supposed significance of randomness in the appearance of new universes would be invalidated.

Creationism and parallel universes
To understand these arguments better, you need to learn more about creationism. People who support this perspective say that chance wasn't a part of how the cosmos came to be. Instead, a creator has provided every physical law and all the factors needed for life. Simply put, everything was carefully planned from the very beginning.

You would be surprised to find out though, that many other religions are not completely incompatible with the meta-universe concept. For example, in one of Buddhism's scriptures, particularly the Jataka Tales, it was mentioned that there are infinite worlds, which are all essentially located between the many levels of heaven and hell.

People who follow the Islamic faith are also familiar with the word "multiverse." The Quran contains verses that talk about the existence of several heavens. In addition to that, the presence of multiple (very similar) worlds is also talked about in the religious text. What makes Islam quite different from Buddhism though, is that a creator's capabilities are considered in the former.

Some experts in Islamic theology argue that the cosmos probably didn't stay as a vacuum since God is more than capable of creating new universes throughout that empty place. Well, given that atoms and other particles were eventually discovered throughout the outer space, it seems that the Islamic multiverse idea isn't farfetched at all.

While often considered dissimilar, Islam and Mormonism are actually quite alike when it comes to the subject of parallel universes. Mormons believe that Jesus created a reality that's different from where God the Father originally resided. While that world isn't the same as ours, the two are supposedly analogous in many ways.

Many Mormons, particularly the religion's leaders, argue that emptiness does not exist. There is always something that occupies an available space. In that sense, it is assumed that the cosmos is not empty. It is being occupied by things that cannot be seen; the space is filled by parallel universes.

Several multiverse theories involve "things that cannot be seen." It is safe to say that religion and science overlap in some ways, at least on the theoretical aspect. What remains to be answered is the question of the existence of an interconnection between these and the world of philosophy.

Think of Philosophy

As you would expect, the field of philosophy is neither about evidence nor faith. It is mostly related to logic. With that in mind, it is obvious that philosophical concepts on the topic of parallel universes do not have the same set of problems as their counterparts in the realms of science and religion. David Lewis' Modal Realism notion stays valid and logical with just a couple of assumptions.

Here is the first of those important principles – there are realities that exist side by side with this one. Given that Modal Realism merely refers to the realness of other universes, that assumption definitely makes sense. Aside from postulating that there are other possible worlds, Lewis made the necessary assertion that any other reality isn't significantly different from ours.

It should be said though, that his idea is not devoid of criticism. For example, there are those who argue that Modal Realism goes against common sense. Even Lewis himself acknowledges this to be true. To be a bit more specific, the multiverse concept often lies within a person's imagination and definitely not within the realm of intuition.

There is also a critique written by Robert Stalnaker that points out Lewis' tendencies to believe in what resides in his imagination. Specifically, Stalnaker disagrees with Lewis' overreliance in analogies in relation to Platonism. This might have made many of Modal Realism's aspects too illogical to be considered valid.

It is true that Lewis' view on parallel universes is not the most fascinating thing that you've read in this book. You might be wondering whether there are other much more intriguing philosophical concepts. There are three main logic-driven notions, and the second one is just about the possibility of identical things appearing in different realities.

The third philosophical perspective, Fictional Realism, is arguably the most interesting of the bunch. It merely refers to the presence or the actual existence of fictional characters. As long as fiction exists in this world, then the entities that reside in those ideas or works also exist within this universe.

By now, you should have already realized that similarities are also present among philosophy, science, and faith in terms of perspectives on parallel universes. Specifically, all three involve numerous assumptions and cannot be considered flawless especially if you don't stick to one particular way of analyzing them.

Can Human Beings Live and Survive on Different Planets?

The Prospect of Life Outside Earth

It is common knowledge that the Earth is already aging. Sooner or later, every life form will perish, and Earth will cease to exist. This is one of the main reasons scientists and experts are considering the possibility of sustaining life outside of the planet Earth. Mars is one of the few planets and stars that many think may be able to support life form.

There have been many expeditions and explorations launched to prove if Mars has some properties that can indeed sustain life and if people can survive there. While more studies and explorations are needed to prove that humans can really live outside of the planet Earth, there is one thing that experts and scientists are hopeful about: human life has the chance to exist outside Earth.

Reasons that Life Exists Outside Earth

As mentioned in the previous chapter, scientists are hopeful that life can truly exist outside Earth. Why? Because there have been theories and evidences that have surfaced throughout the years in support of the existence of life outside our planet. We may know them as aliens, and there are pieces of evidence that prove that they do exist. These are what make scientists and experts hopeful that humans, too, can survive outside the planet Earth. Here are some of them:

Existence of extremophiles – one of the biggest questions people have in mind is if life could exist in a world that is the exact opposite of the Earth. The answer seems to be yes if you consider the fact that even Earth allows extremophiles to exist. Extremophiles are organisms that can withstand extreme weather temperatures, poisonous environments, places with high chemical content, and even in a vacuum. Scientists have discovered organisms that thrive in the extremely hot volcanic dents right at the bottom of the ocean, surviving without oxygen. They have found organisms that thrive in the brackish water of the Andes. There are organisms and animals that live in the extremely cold temperatures of the Arctic. There are even microscopic organisms known as tardigrades that can exist in a vacuum.

These are evidences that life can exist even with an environment exactly opposite of the Earth. Point in case, life can exist in various environments such as extremely hot temperatures like that of the sun and extremely cold temperatures like that of Neptune. It may be just a case of humans not having found it yet.

Chemical precursors found on other planets and moons – many scientists believe that life here on Earth evolved through various chemical reactions that formed cellular membranes and proto-DNA; but these original chemical reactions may have sprung through various complex compounds that are organic like nucleic acids and carbohydrates which came from the atmosphere or even in the ocean. There is much evidence that these compounds termed by many as the "precursors of life" also exist in other planets and moons in the universe. Titan and Orion Nebula are found to have an abundance of these organic compounds though life forms have yet to be discovered. It is believed that these are important ingredients to the formation of life. So if other planets and moons have these key ingredients, it is most likely that life exists in other places, too.

Discovery of Earth-like planets – throughout the years of astronomical explorations and expeditions, scientists and experts have found a vast number of planets and bodies in the universe that exist much like the Earth. Hundreds of these exoplanets are said to be gaseous and giants just like Jupiter. In recent years, through the development of new techniques to detect planets, they have found solid and rocky planets that are very similar to the Earth. Some of these newfound planets are said to be in the "Goldilocks Zone," which means that they orbit in a distance that produces the same temperatures and environment as that of the Earth. With the abundance of exoplanets discovered, it is possible that one or more of them also host some kind of life form.

Tenacity and Diversity of life forms on Earth –scientists are also looking into the tenacity of life forms here on Earth. Many life forms have managed to survive volcanic eruptions, the ice age, meteor strikes, drought, and many other catastrophes. There have also been evidences of a very diverse species on Earth that existed and evolved in a short period of time (speaking in geologic time scale). This tenacity and diversity is one of the reasons scientists think that life form can truly exist in another planet as well.

The mystery of life formation on Earth – how life formed on Earth is still a hotly-debated topic among experts and believers. It is still a great mystery though as to how life truly started to exist on Earth. While some may believe the Big Bang Theory, others believe the book of Genesis. However, none of these theories until this day have shown clear proof that the general public will accept. There is even a theory that said life on Earth began with the chemical reactions between organic compounds. There is still a great mystery as to how the Earth was able to support life considering the fact that the environmental conditions in the beginning of the world was quite hostile. Now, many scientists believe the panspermia theory in which life form here on Earth actually came from and evolved elsewhere – probably Mars – and was brought to Earth through meteorites.

Water bodies can be found in other planets within the solar system – there are evidences that lakes and oceans also exist in other planets within the solar system. It is believed that life on Earth originated from the chemical reactions of organic compounds in the ocean. So it is only logical to think that this may also be true for other planets. There is strong evidence that Mars once had a freely flowing water system and Titan, Saturn's moon, has various seas made of methane, and different rivers that freely flow throughout its surface. Europa, Jupiter's moon, is also believed to be one massive ocean that is warmed by its core and covered completely in a thick layer of ice. Because of the existence of these water bodies, such places may have produced life before or may currently be supporting life forms.

Theory of evolution – though Charles Darwin and his colleagues may not have been aware of exoplanets, the theory of evolution suggests that when a place takes hold of life, it will; and if you consider the environment as interstellar spaces, an interpretation of this theory suggests that life can adapt in outer space.

These are some of the issues scientists and experts are looking into to determine the possibility of life existing outside Earth. There may be more studies and explorations to conduct, but these theories and evidences make them more hopeful that life can exist beyond Earth. Therefore, humans may be able to adapt to a different environment and make it possible to inhabit other planets, stars, and moons as well.

Why Should We Inhabit Other Planets?

It is no secret that scientists and experts are looking into the possibility of living on other planets; but one question that many people harbor is, why should we inhabit other planets in the first place?

One reason could be that in 1989 a small asteroid crossed the orbit of the Earth about 6 hours after the Earth passed the same location. According to experts, the small asteroid, if it collided with our planet, could have an impact equivalent to 1,000 nuclear bombs exploding. According to Lifeboat Foundation, an organization with hundreds of researchers who track various existential risks to humans, said that similar to a game of Russian roulette, 1 in every 300,000 strikes of catastrophe could be it: keep pulling the trigger for a long time and you can blow your head off, and there is no guarantee that the next pull won't be catastrophic.

Most of the threats that made scientists consider living in other planets are mostly due to man-made events. According to World Wildlife Fund, human daily consumption far outstrips what the planet can actually sustain, and by year 2030, humans will be consuming two planets' worth of natural resources every year. According to the Center for Research on the Epidemiology of Disasters reports, the amount of floor, epic rains, droughts and earthquakes we have experienced the past few years is three times the amount experienced in the 1980's and about 54 times experienced in 1901.

This led to most people into believing that Climate Change and Global Warming are the major causes of such events as well as water shortage, famine and coastal areas submerging in the ocean. Furthermore, as what Lifeboat Foundation said, the world could easily end through the misuse of certain technologies such as nuclear weapons. Or, it could be due to a deadly pathogen or through nuclear war. Given the risks human beings pose on the planet, there is a possibility that we leave it if only to conserve what little resources are left. Or, there is also a possibility that we leave it to seek more resources outside.

None of these threats and risks to humans seems far-fetched. The Earth has at one point in time experienced it already – or something similar. Climate Change and Global Warming are already growing concerns within societies and Earth has already experienced a massive extinction of life form due to an asteroid impact. If humans desire to live for hundreds and thousands more years, it is inevitable that humans populate other planets as well. With the constant abuse and misuse the Earth has recently suffered from, it is likely that it will be difficult for natural resources to be sustained – and humans need those resources in order to live and to survive. There may even come a time when there will be more humans living on other planets than on Earth due to limitation in resources.

Prospective Planets

Since there have been more evidences and reasons to believe that life can exist outside of the Earth, the planet is now looking a little less lonely in the universe. NASA has released a statement that there are three planets the size of the Earth that were found orbiting a nearby star with orbits that are ocean-friendly. As mentioned in chapter 2, oceans are where chemical reactions between organic compounds originated that started to form life on Earth. With the presence of ocean-friendly orbits, these three planets discovered have the potential to have an environment that will make it habitable for humans. So, these three planets have the ability to sustain life and allow life forms to survive.

In recent years, astronomers have discovered already more than 800 different planets that orbit stars located nearby. However, only a handful of these 800 planets are the size of the Earth and those that are located in the "Goldilocks Zone" which has the right temperature to make it habitable for humans. Some of these habitable ones are Kepler-62e, Kepler-62f and Kepler-69. These three planets are said to have the closest resemblance to Earth, which can make life form humans sustainable.

Kepler-62e and kepler-62f are said to be the closest to the Earth in terms of astronomical distance. It is about 1,200 light years away from the Earth, which is equivalent to 708,000 trillion miles away. Both planets are smaller than Earth so habitable zones may be closer in. Both planets also have fewer days to complete a year with only 122 days and 267 days respectively.

Kepler-69 is another planet scientists think will be habitable for humans. It is said to be about 70% larger than the size of the Earth and has a year that consists of 242 days.

Aside from the newest Kepler planets discovery, NASA says that humans really have a lot of other planets, stars, and moons to consider inhabiting. One that they are seriously looking into is our very own moon. NASA reported that a colony could be established several feet under the surface of the Earth that has an existing crater to protect humans from constant exposure to high-energy cosmic radiation that can alter the DNA and potentially lead to cancer. The organization even envisions having a nuclear power plant, solar panels, and using various methods to extract carbon, aluminum and silicone for other purposes. It is said that the moon is the most logical place to inhabit due to the presence of ice that is said to be able to sustain life as a catalyst for lunar establishments like hotels, bases, and even casinos.

Though the moon is probably the nearest and most logical place to resettle humans as humans have already landed there, other experts believe that there are other more suitable moons to inhabit such as those of Jupiter, Saturn, Uranus and Neptune. The moons of these planets are said to contain a more abundant source of water, nitrogen, and carbon, which are essential to sustain life.

However, they say that the most similar to Earth and probably the most ideal location to inhabit for humans is Mars. In the latest Mars exploration, it was discovered that there is ice water on the grounds of Mars that is continent-like in size. There are also enough amounts of carbon on the soil to sustain life forms. Temperatures during the day are measured to reach about 70 degrees F. Over time, scientists believe that the planet could be "terraformed" or may be transformed to become like the Earth to make it habitable for humans by using the underground ice water to create an ocean and later be able to create an atmosphere with breathable air for humans as well as establish a shield to protect them from cosmic radiation and other cosmic threats.

However, due to limitations in planet spotting, estimations about these planets can only be given. There is still much to be discovered about these planets to prove that they are indeed habitable for humans and if life form can really survive there. Further research and development in techniques, as well as technology are still in progress, so we can only look forward to future developments to see if it really is possible for humans to live outside Earth. Thus, some experts believe that an alternative may even be more plausible. Instead of inhabiting other planets, launching orbital habitats may be a better idea.

An orbital habitat is engineered in every detail to match the exact specifications required by humans in order to live and survive. Due to limitations in finance and technology, the amount of resources needed to make an orbiting structure habitable is nearly impossible. That very same habitat could be constructed using resources from Earth-like asteroids and astronomical bodies, which in turn can provide more terrestrial variations and provide a larger surface area than all said planets, stars and moons combined. This prospect of humans building an orbital habitat may sound even more appealing because everything that humans need including protection from cosmic radiation and other cosmic threats could be included.

One of the main advantages of an orbital habitat, according to scientists, is that it doesn't need to stay in one orbit forever. If it has already exhausted all the resources of the asteroid or astronomical body it is attached to, or if it needs to run away from a dying star/sun, it could easily be rigged up to be sent to another faraway astronomical body or asteroid through solar sails or nuclear reactors attached to it.

There have been discoveries and studies conducted to find out if living outside Earth is really plausible. Though many planets, stars, and moons have been discovered to be within the habitable zone with some resources needed to sustain life abundantly, the atmosphere and general conditions of the environment still remain a mystery.

How Can We Inhabit Other Planets?

With the discovery of various planets, stars and moons as good candidates to be the next "Earth," questions on how humans can live and survive in other planets have surfaced. Yes, they may have the general characteristics needed as a precursor of life, but how can humans really adapt? How can humans find the resources that they need in order to live there?

The first challenge scientists have to overcome is successfully launching the ship. Some experts believe that once the ship has been successfully launched, nothing is impossible – half the battle is already won. However, the challenge with this is financing. Space shuttles manned by astronauts cost around $450 million per trip. Moreover, sending space probes into outer space without any astronauts already cost $12,000 per pound of weight. It is definitely very costly to do this, as fuel consumed for the first hundred miles can be very expensive. This initial hurdle is one of the many reasons many people doubt the possibility of living outside Earth. This is also a challenge to many scientists to invent rocketless launch systems to make costs lower and make it possible for human beings to afford living outside of the planet Earth.

The next challenge would then be how to find the necessary resources to live. If you have seen Wall-E (the Disney animation movie), it reflects how experts are trying to develop ways on how people can have the necessary resources they need in order to survive. In the case of orbital habitats, the ship or habitat itself could supply the necessary resources such as food, air, water, and other basic needs; but in the case of landing and setting up another Earth in another planet, it would mean extracting a lot of resources from Earth and transport it there to make it more habitable. In this case, there a lot more studies and effort are needed as establishing a new Earth on a different planet is still close to impossible as of the moment.

Another challenge would be on how to enable humans to survive the long trip. As it is many people have motion sickness and we all know that rocket ships are designed to be very fast to negate the pull of gravity. This can be a difficult challenge and a lot of training would be needed for normal people to go on a trip to outer space – just like what astronauts undergo before being launched into space. It may take months or even years for some to get used to the motion before they can safely take the trip to outer space. Additionally, scientists have to develop a ship that can make it more comfortable and more convenient for people to travel in. They need to create a ship that can lessen motion sickness and make it seem like they are simply traveling on an airplane.

How we can inhabit other planets is still a work in progress. There is a lot of research, training and discovery to do for that to happen. This progress may not be in the near future but scientists are hopeful that humans can inhabit other planets when the Earth has become hostile and inhabitable already.

Is it Really Possible to Live Outside Earth?

So the real question is, is it really possible for humans to live outside Earth? In the distant future, it could be a yes. There is much possibility presented by scientists and experts that prove that life outside Earth is plausible. There are hundreds to thousands of exoplanets, stars and moons that have presented themselves as good candidates to sustain life form. These astronomical bodies are believed to have the necessary ingredients to make it habitable for humans and for life to actually exist; but due to limitations in finance and technology as of today, there is still a lot of discovery that had to be done in order to prove 100% that humans can really live outside of our planet.

There is also the possibility that there is currently life outside Earth. Though the necessary ingredients to sustain and create life are there, most of these astronomical bodies do not have the atmosphere to protect humans from cosmic radiation and from asteroids.

Today, we can only be hopeful. Furthermore, it is best that we start caring for our environment today for us to not need to live on another planet anymore.

Is Teleportation Possible?

Introduction

As you may know, teleportation is a popular concept among many science fiction works. Eventually, it has made its way to the world of research through a number of laboratory experiments. To better understand the principles behind teleportation, this book provides a comprehensive definition of teleportation and how it works. It also includes a brief history on how the experiments started. The possibility of human teleportation is also discussed towards the end.

What is Teleportation?

Have you ever wanted to be somewhere else at once without actually having to travel to that other place? Imagine yourself sitting in your office chair, but all you want is to be at home in front of the television or even at a beach lounging under the afternoon sun. Is it really possible to travel to another place without crossing the distance in between? Or could you travel from one point to another without having to spend time and effort in doing so?

The answer, or perhaps the biggest question that these ideas pose to us, is teleportation. In its simplest definition, teleportation allows a person or an object to travel from one place to another without physically occupying the space that is in between. Physics has taught us that the way to travel from one point to another is to traverse the space in between. From point A to point B, there is a certain distance and amount of space in between that needs to be covered before a person or an object can travel between the two points. Travelling requires that the object to be transferred to another location be moving at a certain rate or speed. It also must spend a certain amount of time moving from its origin to its destination.

But what if time and movement are no longer needed? What if you can just think of bringing yourself to your desired destination without you having to physically travel? This is what teleportation is all about. It is about being able to get from point A to point B without having to move through the space and distance separating the two points.

With this possibility, a person can move from one location to the next in just a matter of seconds. One moment you are at home getting ready for work, and then the next moment you are already at your office without having to go traverse roads and deal with traffic. Similarly, any object can be transported without having to physically move through the space between its origin and destination.

The power of teleportation allows you to travel without having to board cars, plains, or any other mode of transportation. You can go wherever you want anytime. You can also send gifts to your friends automatically without having to worry about delivery details or shipping fees. You can bring objects to you without physically getting up from where you are and retrieving them.

Teleportation and Science Fiction

This marvel that breaks down the rules of physics has been in the minds of people from as early as 1931. The term was first coined by American writer Charles Fort. He used it to describe the unexplainable appearances and disappearances of anomalies during that time.

The word teleportation is a result of the combination of the Greek prefix *tele*, which means "distant," and the Latin word *portare* meaning "to carry." Unfortunately, the term and the whole concept seemed to be a result of superstitions, lies, and even some hoaxes. It appeared, at some point, to simply indicate how inadequate science is in explaining such strange phenomena.

While the term may have come across a lot of criticisms in the world of science, it was greatly welcomed in the world of science fiction. One of the most memorable icons of science fiction that makes use of the idea of teleportation is Star Trek. In the story, Captain Kirk and his crew would beam themselves to and from foreign planets with the use of a teleportation device. They would easily travel from their ship to the surface of the new planet, and they would also be able to beam themselves back up to the safety of their ship.

In the world of superheroes, teleportation is also a very welcome superpower that can be used both as offense and defense. Mutants and superheroes are able to move to and from different places without having to physically travel. Most of the time, they just have to think of the place they want to go to, and they can be there in just a blink of an eye. In most cases, they are also able to teleport other people or objects with them.

While the normal idea about teleportation is simply for it to be used as a means of travel, it can also be a very powerful form of offense. In cases where the teleporter is able to move quickly from one place to another, it can be used to mount surprise attacks on an opponent. Even more useful is the ability to teleport other objects to use them as a form of offense. For example, teleportation can be used to send objects that would crash into the opponent. A teleporter can even transport the opponent himself within the range of danger. Another tactic is for the opponent's attack to actually be teleported back to him so that it backfires and causes damage upon himself.

The power to teleport holds so much potential that it has even given people a glimpse of the future. Many works of science fiction relate teleportation to time travel because of the person's ability to travel through space as well as time. If we look at teleportation being defined as a person or an object disappearing from one place and appearing in another, the concept of time travel is not at all far from it. This is because time travelling involves the disappearance of a person from his current time and his reappearance at a different point in history. With this, time travelling is sometimes considered as a more complex form of teleportation.

In most cases, time travel makes use of a special machine or equipment that serves both as a teleportation device and a time travelling device. The use of such teleportation vessels is also common across works of science fiction. Normally made up of a pair of these machines, a teleportation device is set up so that one device is placed where the teleporter currently is and the other device is located where the teleporter wishes to be transferred to. While this is not as powerful a teleportation as the ones that superheroes are imagined to have, these are actually more believable in the sense that people think that these teleportation machines can actually be made. It is quite hard to imagine that a human being can transport himself or any object to any location, but there could be a possibility that a device would be invented for such purposes.

With the possibilities that teleportation offers, there are many who actually try to find ways to acquire such powers. Tests and theories are being carried out to see if teleportation can in fact be done. After all, this type of travel will not only make life so much easier, but it can also be a venue for more discoveries and possibilities for the human race.

How Teleportation Works

Many people, especially science fiction fans, are interested in human teleportation because it is such a promising concept. In principle, teleporting humans may be possible, but the chances of it actually happening are very slim. However, the teleportation of smaller objects has been proven possible by many laboratory experiments.

You already know that teleportation eliminates time and space from the whole travel process. Basically, what happens in teleportation is that an object becomes dematerialized from its source location and then its exact atomic configuration is sent and reconstructed in its destination. What this means is that all of the information of the object to be teleported is extracted through scanning. This exact information is the one used by the receiving location to construct a copy of the original object. This replica does not necessarily come from the original object's actual material, but it is perhaps made up of the same kinds of atoms arranged in the very same pattern as its source object.

Teleportation devices work very much like fax machines. It does not send duplicate documents, but is used for 3-dimensional objects. Such device would give out an exact copy of the object to be teleported instead of just an approximate facsimile; however, this object will most likely be destroyed during the process. In many science fiction works, the original body of the teleporter is usually preserved and the plot gets tangled up when both versions (original and teleported) of the same person eventually meet within the story. Of course, the more common version of the device is the one that destroys the original, and it works as a super transportation machine instead of a perfect replicator of objects and bodies.

With these things in mind, it is then worth noting that teleportation does not really move the object, as there is no mass that moves from one location to another. What is actually moving is the information that the particular object holds.

To illustrate this concept, imagine that you have two identical atoms that are in Point A and Point B, respectively. The atom in Point A is in a certain state, and the goal is to move that state over to Point B without destroying it, touching it, or even looking at it. Through teleportation, the essence of the atom from Point A is moved or reconstructed in Point B, but the atom itself never leaves its original location. For this to work, the distance between the two points also cannot be that far apart -- just a couple of meters, to give you an idea.

Although the aforementioned principles do not sound very much like the teleportation that are seen in films and books, those are basically what happens when an object is being teleported with the help of modern technology. When an exact copy of the original object is made at its destination point, it is basically occupying two spaces at once -- which is *here* and *there*. In order to complete the entire teleportation process, the original object needs to be destroyed so that it doesn't stay *here*. Instead, it will be only *there* and only one copy of the object will remain.

At this time, teleporting solid materials are still unlikely, but experts say that teleporting a single atom is possible in theory. Depending on the available technology, perhaps we can eventually teleport groups of atoms and other more complex objects. However, scientists doubt that we will ever get to that point. For them, the whole idea of teleporting particles may prove to be useful only in computers and communications technology.

Teleportation Experiments

We were first introduced to the idea of teleportation when Star Trek hit the television screens during the late 1960s, but it was in 1993 when it first made its way to the world of research and scientific theories. During this time, physicist Charles Bennett and a research team at IBM discovered that quantum teleportation was indeed possible. In the process, however, the original object to be teleported would be destroyed. This discovery was first revealed by Bennett in March 1993 at the American Physical Society's annual meeting. Ever since then, many experiments involving photons have confirmed that quantum teleportation is possible.

In 1998, a group of physicists from Caltech (California Institute of Technology) together with two European research groups tested the IBM team's ideas. They were able to successfully transport a photon, an energy particle that carries light, by reading its atomic structure and sending its information across a meter of coaxial cable and creating a copy of it. As Bennett's research suggested, the original photon ceased to exist when the replica was produced.

One of the main barriers to teleportation is the Heisenberg Uncertainty Principle, which states that the location and speed of a particle cannot be known at the same time. The main question then was how can you teleport an object if you can't know its exact position? To proceed with the experiment without going against the Heisenberg Uncertainty Principle, the research group from Caltech made use of a phenomenon called "entanglement." At least 3 photons were required for this work:

Photon A, or the photon that is to be teleported;

Photon B, also known as the transporting photon; and

Photon C, or the photon entangled with Photon B.

The reason why there were three photons was because the researchers could not closely examine Photon A without entanglement, as there is the tendency to disturb it and therefore change it entirely. The two entangled photons, B and C, could obtain information about Photon A and these would be passed on to Photon B through entanglement, which would then be transported to Photon C later on. This whole scenario enabled the researchers to apply Photon A's information onto Photon C, creating an exact replica of it. As a result, Photon A would no longer exist by the time its information made its way to Photon C.

In 2002, a research group from the Australian National University teleported a laser beam with success. As of now, the latest successful experiment in teleportation occurred on October 4, 2006 in Denmark. It was at the Niels Bohr Institute in Copenhagen where Dr. Eugene Polzik and his research team teleported a laser beam with stored information into a cloud of atoms. For Dr. Polzik, it was indeed a milestone because it was the first time that a teleportation experiment was conducted between two different objects such as light and matter, where one is the storage medium and the other is the information carrier. The stored information was teleported approximately half a meter away from its source point.

Most of these experiments hold a big promise for the world of quantum computing. This is because the results from these studies can contribute significantly to the development of various networks that will make the distribution of quantum information much easier. One day, this kind of technology may be utilized to create a quantum computer that can transmit data at a much faster rate than the most powerful computers that we have right now.

Human Teleportation

With all these experiments that prove that teleportation can indeed happen in reality, we move on to the big question: is human teleportation possible? After all, if scientists are now able to teleport photons and atoms, surely teleporting bigger objects –– ultimately a human being –– can also be possible.

Let us look at where science is now with teleportation and how much farther it is from human teleportation. So far, the concept of teleportation is made into reality with the use of replication. Data is sent from one photon to another, and this data is used to make an exact replica of the first photon. In essence, the original photon is not exactly transported to its new destination, but rather, an exact replica of it is created in a different location. The question now lies on how possible it is to do the same thing with objects that are millions, even billions of times larger. Photons and atoms are invisible to the naked eye, and are so small that the relevance of being able to teleport them into new locations seems insignificant to the normal person. But what if you do the same thing with a visible, tangible object?

Numbers and Possibilities

Let us look at the paper clip for example. It is visible to the eye, very tangible, and having it wherever you want it can be useful as well. Now, a single piece of steel paperclip actually contains at least a trillion atoms of iron and carbon. This means that what is needed to teleport one single atom needs to be multiplied at least a trillion times over just to teleport a single piece of paperclip. The mere number makes this probability far from where science is today, but the success of previous experiments shows that it is possible.

Now, with a human person, the number of atoms is even larger. Compared to a single paperclip, the human body is made up of at least a billion times more atoms. What makes human teleportation even more complex is the fact that the human body is made up of different kinds of atoms, including oxygen, hydrogen, sulfur, calcium, and many more. All of these atoms have to be treated in a special way for an exact replica to be produced. Also, the atoms that make up the human body are arranged in a very specific way, and that's what makes a particular person unique and different from all other persons. If the order and position of any of the trillions of atoms are changed in any way, the result would not be the same person or object that is being teleported.

Computations

Supposing that all those big numbers are not an issue and that the teleportation process has been mastered so much so that there would be no errors in replicating the human being, another huge factor would be just how much resource would be needed to carry out the full process. This includes the energy, the computing power, the memory space required to hold all the information, as well as the time it will take to carry this information and move it to another location.

In terms of computing power, it is estimated that 10^{23} bits are needed to record all the details of every atom that make up the human body. A machine with such computing power is yet to be developed. Even if it did exist, there are still a lot of other factors to consider. For example, the time that it would take to transfer all of that data and information would be a very long period indeed. This is because there is just too much data to analyze. The transfer will also depend on the computing capability of your machine.

All the data that needs to be transferred to teleport a human being would be roughly around 2×10^{42}. Even with a very high-speed processor, this would still take a long time. For example, a computer that has a computing speed of 30GHz would still take more than 4×10^{15} years to complete the desired human teleportation process. To grasp this number better, let us just say that it is thousands of times larger than the age of the earth, which is currently at 4.54 billion years.

Of course, these limitations may be countered with advancements in technology and science. As of now, however, the numbers do not support the possibility of human teleportation in the near future.

Teleportation: Appearances and Disappearances

The mere number of atoms, as well as the seemingly infinite number of possibilities of how these atoms can be arranged, makes human teleportation far more difficult and complex than the teleportation that science is now able to achieve with photons and singular atoms. But the real problem with human teleportation is in the fact that the teleportation that science is capable of executing right now is the kind of teleportation where only a replica of the original atom or object is produced.

If we look at teleportation being defined as an object appearing in a new location and disappearing from its old position, we are then left with the question: what happens to the original object being transported? If the goal is simply to send a replicate of the object to a new location, then there would be no problems except for the fact that every teleportation process would result to another unit of the object being produced. That is, if atom A is teleported, the atom C that is now in the new location is exactly the same as atom A. In essence, there would now be 2 atom A's. Repeat the teleportation process and this would result to another atom A, and so on. Now, if you were a person trying to take over the world, then this would not be such a bad idea at all.

The definition of teleportation, however, still requires the disappearance of the person or object from its original position. As of yet, the only sensible way to make the original object disappear would be to destroy it. If it were a human person being teleported, then this would mean that the person, the original that is, would have to die and be erased off the face of this earth. It does not take a lot of thinking to know that this is not a good idea at all.

While it may be possible that an exact replica of the person, including all of the personality, the traits, and even the memories, is made, would it justify the death or the annihilation of the original human being? If you were the person to be teleported and it was explained to you that you would die but still technically exist in the process, would you approve of it?

Not only will this be difficult for the person to decide on, but this will also raise a lot of issues including religious, political, and ethical ones. In essence, it will not be a question of whether human teleportation is possible or not, but rather, will people allow it or disagree with it. More importantly, will this be considered a moral act?

To Teleport or Not to Teleport

The possibility of teleporting a tangible object or a human being may be light-years away from now, but it is still undeniable that teleportation is a truly amazing concept. Travelling to places all over the world would be such a breeze and time would no longer be an issue. There is also the benefit of not using up the world's energy resources such as fuel and gas. Additionally, teleportation as the means of transportation would mean less carbon emissions and other forms of wastes, thus preserving the earth and protecting the environment.

It is easy to see how many people would prefer such a form of transportation. There would be less anger due to traffic jams, fewer headaches brought on by delayed flights and trips, and less stress from having to deal with the everyday commute that is part of daily life. It would also be a lot safer as causes of accidents can be avoided.

Indeed, if teleportation was mastered and its efficiency was maximized, there would be no doubt that everyone would be in on the teleportation wagon.
However, with the path that science is on currently, it is hard to say that the teleportation that we appear to be capable of is ideal. Again we go back to the simple fact that teleportation experiments were able to produce exact copies or replicas of atoms and place them in a different location from where the original object is. Unfortunately, it is not the object per se that has been relocated, but merely the information required to create the replica.

If we take a step back and look at what teleportation really is, we will see that we have deviated from the teleportation that we have found so desirable. After all, many people will define teleportation as transporting an object to a different place almost instantaneously and without the need to travel the space and distance in between.

No copies or replicas are created and definitely nothing will have to be destroyed. If this type of teleportation were made possible by science, then it would definitely be the transportation mode of choice. You can travel around the world and transport objects and various items to different locations. Even space travel could be made easier if teleportation were mastered and maximized.

For now however, we will have to be satisfied with the "cloning" type of teleportation that current science and technology is capable of. We will also have to bear with the problems that this type of teleportation brings. But with such problems, it would be much easier to accept life as it is and simply choose to travel the old fashioned way. Walk, ride a bus, get on a plane, or even board a ship. All of these choices will have you spending a lot of time, energy, and even some of your hard-earned money, but they are still the more practical and proven safe methods for getting yourself from one place to another.

Is Time Travel Possible in Our Universe?

Fascinating Stories

Since when did you become interested in time travel? Did you experience something so painful that you want to jump back a few years? Is a past mistake making you wish that you could go back in time and do things over? Maybe you've just been fascinated by the idea of cracking the mysteries of time since you were young. Whatever your reason might be, you have the power to satisfy your curiosity.

The best way to begin your journey towards a thorough understanding (when it comes to the topic of time travel, that is), is to tackle this interesting question: are there people who've already managed to jump years into the future or into the past? The answer to that could actually be yes, depending on how much of a skeptic you are.

Here's one of the most fascinating stories that seems to indicate that folks from the future may have already visited ancient worlds – Chinese archeologists discovered a small watch-like object (probably a uniquely-designed ring) inside a several-centuries-old coffin that supposedly hadn't been opened before. Unfortunately, there's no way to check whether they're telling the truth or not.

It's safe to say that the tomb-and-watch story is a bit hard to believe. Those who've made the discovery could very well be the ones who put the timepiece-like accessory into the casket. Well, this time-travel tale should be a bit more believable – back in the early 1930s, Bernard Hutton and Joachim Brandt were asked by their boss to come up with a report on Hamburg's shipyards.

After they've gathered all the info that they needed, they finally decided to go home. Instead of being able to do just that, they came face to face with the threat of enemy attack. They ended up in the middle of what looks like a British invasion. Here's what made their experience truly perplexing though, the siege suddenly ended and the pictures that they took didn't even match what they saw.

As you've probably guessed, the German shipyards were bombed more than a decade after Hutton and Brandt's experience (or more specifically, during the height of the 2nd World War). If you're not into unexplained wartime occurrences and you prefer events that are much more recent, then you'll love everything about the mystery surrounding John Titor's sudden appearance (on the internet).

What's so special about him? He claimed to be someone from 2036. While he could merely be another netizen who's looking to have fun online, From 200 to 2001, John Titor divulged all kinds of information supposedly from the future. Among those details, one really stood out – he said that CERN (a global physics organization) would do research related to black holes, which in turn would be the basis for time travel.

Was there any truth to what he said? The answer to that is yes. CERN did – in part – become involved in studies about the outer space's swirling abyss. While the knowledge acquired from those scientific endeavors hasn't been used so far to gain insights on traveling through time (or at least, that's what the public has been told), it's still fascinating that Titor was already aware of CERN's pursuits back then.

There are many other stories about people who've moved to different points in history (whether accidental or intended). It's true though, that many of those narratives can't either be proven or debunked, especially since it's virtually impossible to gather more details about them. Nonetheless, the existence of time-travel tales is a good sign for the hopeful.

Mystics' Perspective

Now that you have realized just how mysterious time travel is, you're most likely wondering whether it's related to mysticism. Well, many mystics dwell on the topic of time, time travel in particular. Some of those who follow Jewish beliefs, for example, claim that time-travel isn't only possible – it's something that's very much related to human consciousness.

Once a person begins to perceive the world without relying on the Ruach (or in other words, the logic-driven mind) and shifts to a soul-based awareness, then moving through time or space should be easy. Why's that? The soul works in the world of Beriya – a plane where form isn't relevant anymore since everything's essentially composed of divine energy.

Did that confuse you? If it did, then you only have to remember that logic makes humans focus on limitations. Once they realize that things can actually exist from nothingness, they'll finally be able to grasp notions that normally defy common sense (such as the intertwined nature of time). Now that you have found the answer to "what?" and "why?," you're thinking about the "how?"

Gaining access to the world of Beriya isn't easy. Only those who've strengthened their faith and exerted effort to meditate whenever possible could reach new levels of awareness. If you believe that meditating is effortless, then you really have to keep this in mind – there's more than one kind of meditation (those that involve images and the ones that affect emotions are just a few examples).

Did you say that you're not that keen about delving into Jewish traditions? You don't have to worry at all. There are mystics from other religions who've pondered deeply about time. Angelus Silesius is among them. He's a Roman Catholic priest who lived during the

1600s. Unlike many other servants of the church, he thought that humans have the power to manipulate time.

It should be pointed out though, that he didn't really talk about the possibility of visiting the future. He didn't even mention anything about past-changing opportunities. What he focused on was this concept – time exists due to the mind. He even argued that time only stops when the mind ceases to function (or in other words, when a person dies).

So, what does that have to do with your dreams of becoming a time traveler? Just think about this – if the flow of time is merely something that's made by the mind, then the psyche could actually be the key to traversing the past and the future. In a way, Silesius believed that people could slow down or even stop time if they would concentrate hard enough.

"How about Islam?" This is a question that's on your mind, right? After all, it is among the largest religions in the world. Interestingly, those who follow the Islamic faith might argue against the likelihood of traveling through time. The reason for that is quite simple – if you think about it, time travel could be used as a way to achieve immortality (Islam teaches that everything eventually ends).

At this point, you're finally aware that just like the general public, mystics (or even those who merely consider themselves as religious), have different opinions when it comes to instantly moving across eras. Some are confident that time travel is something that's far from impossible. Others know for sure that jumping into the past (or into the future) can never be done.

All about Paradoxes

What do you think would happen if you were shot in the back by the younger you? In a way, that could be your ticket to immortality (remember the Islamic view?). Your younger self would continue to live your life and eventually die the same way you did. Here's the problem with that – you're actually just creating a loop. Your new self wouldn't have the knowledge and experience of the old you.

Forget about Jewish concepts for now and look at things through a philosophical lens. You have to focus on logical flaws (or time-travel paradoxes as many call them). Here's one of the most popular examples of those – the Grandfather Paradox. What is it about? If a person travels to the past to meet a certain goal, then the reason for the trip would be gone once the objective has been reached.

If you're wondering why that would make the entire concept of time travel (or in particular, going back in time) questionable, then you need to consider this scenario – you want to stop someone from crossing the street to prevent an event that already ended a life once, and you managed to do just that soon enough. The moment you save that person's life though, your goal ceases to exist.

Specifically, given that the future is shaped by past events, then you'd no longer have a reason to travel through time. So, what happens then? Essentially, time travel would be removed from the equation, which in turn means that it's essentially impossible for it to exist. As you'd expect, there are experts in philosophy who've been thinking about paradox workarounds.

Here's an example of those alternative concepts – the consistent causal loop, which is based on the assumption that time (including the past, present, and future) flows on a grand scale. If you do something that involves time travel (such as stealing something from the future so that you'd be famous in the present), then you haven't gone against the normal flow of things.

That's right; the person you stole from would have been your future self all along. While that notion doesn't reflect the event-altering power of time travel (which is most often emphasized in works of fiction), it doesn't invalidate the need to go beyond the present, either. Well, some modern-day philosophers believe that time doesn't work that way. That's why the post-selected model exists.

That odd-sounding term pertains to time's change-preventing powers. For example, if you jump a few years back to change your present and future, you won't be able to do anything, no matter how hard you try. Why's that? Things would always get in your way. To make the idea easier to comprehend, it's best to use the previous example (saving a life) once more.

When you've finally arrived in the past and you're ready to shove someone away from harm, you might end up being accidentally tackled by another person or slipping before you managed to do anything – the possibilities are endless. While this idea doesn't necessarily make time travel seem a lot less achievable (logic-wise), it does make time machines (if ever one would be invented) seem a lot less useful.

So, now that you've finally reached the end of this chapter, you should be able to understand one thing – from a philosophical standpoint, time is much more powerful than human will. After all, even if time travel were to become possible, it wouldn't do much good due to either one of these factors – the existence of a grand flow of things or time's own change-preventing mechanism.

Physicists' Solution

So far, you've become aware that mystics and philosophers don't really see eye to eye when it comes to the topic of time travel. Well, even among those who follow the same kind of logic or faith, disagreements about that matter are abundant. If you're thinking that scientists are any different from those people, then you're completely wrong.

Some of the brightest minds in physics claim that traveling through time is something that would most likely remain in the world of fiction. Stephen Hawking for example, argues (although a bit jokingly) that the lack of tourists who've come from the future is among the greatest evidences against humanity's capability to eventually build a time machine.

Here's what you should know though – he's aware that there's a flaw in that idea. There is a chance that the people of the future know how to travel through time, but they're hesitant to do so (especially for trivial goals). He also admits that there's another way to explain the lack of time-traveling tourists – the future hasn't reached a point where a fully functional time machine already exists.

While Stephen Hawking is a known skeptic when it comes to anything that involves time travel, Carl Sagan (an accomplished astronomer and novelist) is a firm believer. To be a bit more specific, he doesn't dismiss the possibility that those from the future have already explored the present. It's just that they might cause more harm than good if they reveal that they're from another age.

This discussion wouldn't be complete without mentioning anything that's related to Albert Einstein. Truth be told, he didn't pay much attention to time travel (particularly in terms of how it could be achieved or what it might be used for). However, he did come up with a very interesting theory – Special Relativity. That concept simply points out that time and space function together.

If those two things work as one (and couldn't be separated from each other), then anyone who moves quickly enough should be able to age much more slowly. Here's an easy-to-understand example – someone who manages to travel at the speed of light ages at a pace of less than ten percent (compared to a person who's moving at normal speed).

Einstein came up with another theory that's frequently been associated with time travel. That scientific notion is none other than General Relativity, which merely emphasizes that gravity is among the many aspects of space and time (once again in a unified sense). It should be mentioned though, that the iconic physicist wasn't the one who discovered the link between that idea and time travel.

Kurt Gödel was among those who exerted effort experimenting with Einstein's Field Equations. Unlike many scientific experts who engaged in the same endeavor, he discovered something that's really unique – a space-time model that has timelines, which move outwards but eventually curve back to their starting points.

Anyone who's in that space-time setup should be able to jump back several years by merely following the curve. As you'd expect though, Gödel's scientific achievement doesn't refer to something that actually exists. Aside from that, he picked a very specific numerical constant to complete his model. So, his Field-Equation solution only serves one purpose, and it's to point out a possibility.

Technology Secrets

You're most likely thinking of one question right now: did anyone try to make a time machine yet? After all, no matter how fascinating those concepts might be, they would all go to waste if they weren't used for technological progress. Well, there are scientists who've been thinking about building a contraption that could traverse a wormhole (otherwise known as a warped region of space and time).

As you would have guessed, they didn't pursue the idea (or at least, they're nowhere past the planning phase). The reason for that is this – wormholes work in a way that automatically synchronizes everything that they connect. In other words, even if a wormhole connects 1999 and 2013, anyone who enters its 2013 end would just exit through its 1999 end but fourteen years later.

Obviously, that isn't very helpful (unless the many-worlds theory is taken into account, which is completely another matter). Fortunately, there's a time-travel device that sounds a lot more promising – the Tipler Cylinder. By spinning at very high speeds (probably near the speed of light), that device moves anything that passes by it into different points of time.

Here's something that you should know, though – while the machine supposedly functions in a straightforward manner, there's no material that could be used to build it. Simply put, every known substance would surely disintegrate if spun at almost 300 million meters per second. By the way, there's another reason why it's currently impossible to build a Tipler Cylinder – propulsion limitations.

If you don't know what that means, then just think about this – even the most powerful propulsion systems could only accelerate an object several times past the speed of sound (surpassing speeds of more than 3,000 meters per second is already considered a remarkable feat). So, it's safe to say that it'll take centuries before that time machine plan becomes a reality.

It's alright if you're beginning to wonder whether there's a chance that you'd see an actual time-bending contraption in your lifetime.

Well, two physicists from Vanderbilt University (Tom Weiler and his colleague, Chui Man Ho) believe that the largest particle collider could become the very first time machine, depending on whether a successfully isolated Higgs-Boson splits into smaller units.

Why would that be relevant to time travel? When that particle splits, it supposedly crates Higgs Singlets – minute things that can pass into another dimension that's connected to another point in time. Even if Weiler and Ho are correct though, knowing that particles have traveled to another point in history doesn't mean that people would be able to ride colliders into the future.

The closest things to a time machine that humanity has built so far are future-predicting devices. No, this isn't about gadgets that give zodiac analyses and tarot readings. These are contraptions that use all sorts of info that have been gathered from the past and the present to assess whether certain events would occur in the future. The Aryayek Time Traveling Machine is an example of those tools.

Now that you're fully aware that technology still hasn't reached a point where it's capable of sending people into different time periods, do you think that time travel is something that could ever be achieved (from a scientific point of view, that is)? Are you beginning to believe that changing things in the past or in the future could only be done in fictional worlds?

Imagination Power

There's a chance that time and travel are words that would only work well together in fantasies. While that might be a disappointing answer to "is time travel really possible?", people might actually be united by their yearning to either change the past or affect the future. If you're wondering how time-travel could serve as a unifying factor, then you need to ponder upon literature.

Back in 1895, H.G. Wells' finest literary creation (*The Time Machine*) became a hit for one simple but important reason. It was the first book that contained the term "time machine". Even though time travel isn't a very fresh concept back in those days, the public couldn't help but be fascinated by the possibility of riding a contraption that could be used to explore different periods.

Did that confuse you? This should make things much clearer – some ancient works of fiction already involved characters that jumped into the past or even leapt into the future. Only the notion of having some sort of time-traveling equipment was completely unheard of (control is the keyword here). What's that? Do you want some examples of those antique writings?

Well, here's one that should intrigue you – *Urashima Taro*. That Japanese folk tale is about a fisherman who found a grand, thriving palace underwater and chose to stay there for a couple of days (as a visitor, of course). When he finally decided to return home, he was shocked to find out that everything had already changed because 300 years had passed on the surface world.

Although not as ancient as that of the Japanese story, the French also have something to be proud of when it comes to time-travel literature. Back in the late 1800s, one of Pierre Boitard's masterpieces (*Paris Before Men*) was published. What made the story stand out is that it revolved around a person who unwillingly ended up in the prehistoric age.

As you would've guessed, the tale had dinosaurs and ape-like creatures in it (Boitard wasn't really a novelist by the way, especially since he was mainly interested in the sciences of rocks and plants).

After learning about those fictional adventures, have you finally realized why time travel seems to connect people throughout the world?

That's right; it's safe to say that individuals of different cultures all have their very own take on how tough (yet breathtaking) it would be to have a period-jumping experience. Being curious of time-related possibilities seems to transcend both history and race. Well, here's something that you should know – the world still can't get enough of era-traversing tales.

New fictional narratives are still appearing one after another. In 2006 for example, a movie (titled *Déjà Vu*) tackled the interconnected nature of time (or how moving several years back could actually be the reason for the present to be shaped as it is). To be a bit more specific, the film was about an enforcer who ended up killing his partner by trying to send a life-saving message to the past.

On the other hand, *Looper* is a 2012 movie that's about achieving immortality by traveling to an earlier period (you already know how that works, right?). All in all, it's only to be expected that Hollywood (and filmmaking industries in other regions of the globe) would continue to come up with stories about time travel – that is, until someone finally invents a real time machine.

Are UFO Pilots Really Aliens?

Introduction

UFOs and aliens are the sci-fi versions of ghosts and spirits. Spotting any of them is a rare opportunity; and when you see one, few will believe you while others will think that you are either mentally ill or high on drugs. The biggest difference between the extraterrestrial and paranormal beings is that science has the possibility to prove of their existence.

Nowadays, people are skeptical about reports regarding unidentified flying objects. Only those who are UFO fanatics are all agog when a new UFO sighting is reported on television. Experts debunking multiple sightings of UFOs and people creating alien stories caused the decline of interest about extraterrestrial beings.

With the advent of modern technology, some rarely believe in the likelihood of aliens being involved in UFO sightings. They have the prejudice that most of the UFOs that people see are mostly manmade. With high-end cameras at the disposal of many, low-resolution UFO footages have become preposterous and suspicious. On the other hand, with the existence of excellent photo and video editing applications, one can simply forge a UFO sighting.

On the other hand, almost half of regular people and UFO fanatics strongly consider that those things might be related to deep conspiracies planned by government and shadowy institutions. With a government that is a bit questionable in every aspect, who would not resist thinking about the possibility of a conspiracy? In addition, with some of the sightings supposedly happening relatively near some military installations, who would not think that there is something fishy going on?

Unfortunately, the only ones who can disprove or prove those theories are the UFO pilots themselves. However, is it possible for humans to get a hold of one? If one will be able to see the pilot, who will it be; or the more appropriate question here is – what will it be?

Are UFO pilots aliens?

Or are they humans?

Also, the possibility of being unmanned is high.

In case you are looking for the answers to those questions, this is the book for you. This book will be able to provide you some hints. However, remember that this will not give you conclusive answers. Take not that you are treading in an unknown territory – the existence of terrestrial beings. Many scientists have already given up on their search for the answers that you seek; and they had already reached a point where they already considered these things are pseudo-science. Nonetheless, there are a few things that will happen to you after you read this book. You will definitely ask more questions, gain more curiosity, and starve for more answers.

The Possible Extraterrestrial Pilots

UFOs or Unidentified Flying Objects are mysterious things found floating, flying, hovering, or soaring in the sky. They are mostly associated with flying saucers and aliens. However, some people think that these objects are handiworks of the government. Usually, radars, telescopes, cameras, and human eyes can spot UFOs.

The items in the list below are the potential alien beings that may have piloted the UFOs that humans have seen so far. The existence of all of these beings has been testified to be true, and incidents surrounding them are very common. That is especially true during the UFO craze that happened a few decades ago.

On the other hand, most of the incidents were related to UFO sightings, and were reported by people who experienced close encounters with the beings. Together with the information regarding their appearance and behavior, their involvement with UFO incidents will be mentioned in the list.

Andromedans

Andromedans are said to be aliens or extraterrestrial beings from the Andromeda constellation. An American citizen who used the alias Alex Collier announced the existence of such beings. He claimed he was a 'contactee' for the Andromedans; and according to him, Andromedans are Nordic aliens.

The Andromedans told him that he needed to relay some of the information that they know to the human race. What made people believe his statements about those aliens are the secret histories that were relayed to him. Most of the stories are about alien colonization on Earth, extraterrestrial conspiracies, alien classifications, and ufology practices and traditions of humans. In addition to those stories, he claimed that he was forced to remain unknown because of the threats he received from three men who were all 'well-dressed'.

The message he needed to pass on was about the opportunity of the human race to transcend to 'higher realms of reality'. To do that, human beings need to eliminate chaos and harmonize before August 2001. He referred to the message as 'vibrational densities'.

Flatwoods Monster

The Phantom of Flatwoods, Braxton County Monster, or Flatwoods Monster is another extraterrestrial being that was allegedly seen in Flatwoods, a town in Braxton County, West Virginia. The first alleged sighting of the entity was on September 12, 1952.

According to the testimonies of those who witnessed its existence, the Flatwoods Monster is three meters tall, has a greenish body, and has a reddish face that glows. The creature's head has protruding eyes that are heart shaped. In terms of structure, its body proportionally resembles an ordinary human being. However, a few claimed that it does not have any arms while some said that it has broad, short arms and has hands with claws.

Aside from those details, eyewitnesses mentioned that a red glowing ball-like entity hovers above its head. They also mentioned that the glowing ball rested on the ground sometimes. On the other hand, ufologists claimed that the glowing thing was actually a hovering craft that the Flatwoods Monster piloted.

Greys

In contrast to the green colored body of the Flatwoods Monster, Greys have dark grey skin – hence they were named after the color. Another feature that separates the Greys from the Flatwoods Monster is their size.

Compared to humans, they are smaller. Most of the reports regarding the sightings of the Greys mention that they are estimated to be two to four feet tall. Nevertheless, despite their small stature, they look tall since their bodies are vertically elongated. However, their upper and lower body is disproportional. Few of the noticeable disproportions in their body structure are their small chest and boneless appearance. Even if their chests are small, their upper bodies are more than half the length of their lower bodies.

Also, they do not have external body organs like ears, noses, and sex organs. Instead, they have holes as replacement to those 'missing' organs. They do not have any visible hair, and their face is top-heavy, too. When it comes to their eyes, they have no discernable pupils or irises.

Hairy Dwarfs

Hairy dwarfs look like diminutive humanoids – just like the Greys. The main contrasting feature against the latter is their hair. Just to help you visualize its appearance, a Hairy Dwarf shares the same appearance of Bigfoot, though they are shorter and are more adept in science and technology. Most of the reports that were gathered regarding Hairy Dwarfs mention that these extraterrestrial beings use tools, wore cloths, and have the ability to speak. On a different note, some have witnessed Hairy Dwarfs that are taller than humans. Those large varieties are often compared to werewolves in terms of physical characteristics.

The number of sightings of Hairy Dwarfs was relatively high during the early days of the UFO craze. Nowadays, reports about them have become uncommon, while the sightings that include UFOs and Greys have increased. Back then, whenever a Hairy Dwarf was spotted, UFO researchers believed that it signified a possible UFO visit.

Hopkinsville Goblins

The Hopkinsville Goblins are three feet tall, have claws, and pointy ears. Their skin is said to be metallic. Also, their limbs are very thin, and their legs are very skinny. Apparently, they are strong enough to endure and survive gunshots from shotguns and rifles. During the Kelly-Hopkinsville encounter, the men who were around that time tried to shoot the Goblins at close range; and whenever they were hit, a metallic rattling or ring noise resonated from their body. Aside from that, the beings were capable of floating.

The Hopkinsville incident is one of the most convincing UFO stories up to now. The case happened one night in 1955. With 11 people, composed mainly of friends and family members, they sighted numerous goblins wandering outside the house. Apparently, not one of them was able to get inside the house. According to the people inside the house, the extraterrestrial beings created clawing and scratching sounds outside. Aside from the accounts of the 11 people, a state trooper also testified that he heard the same noises at the same time the incident was supposed to be happening.

Most of the fictional stories that involved aliens borrowed ideas from this case. On the other hand, it was difficult to dispute the authenticity of that event due to the number of witnesses, lack of reason to fabricate the story, and the consistency of the testimonies from the witnesses. To add to that, one week before the Kelly-Hopkinsville case, the same incident happened in Kentucky. That incident was also difficult to disprove due to the number of eyewitnesses.

Little Green Men

The most stereotypical look of aliens was derived from Little Green Men. Fundamentally, they look like gremlins with antennas on top of their heads. They are commonly associated with the Greys due to their many similarities – the only obvious difference between them is their skin color.

Believable sighting incidents of Little Green Men are very few; and many people think that most of the events that include the said creature were just figments of the witnesses' imagination. It is a fact that the name or term 'Little Green Men' originated from folklore. Many novelists and people use this term for supernatural beings – a good example is the gremlin. Since that is the case, many people were influenced into believing that aliens are actually Little Green Men. Because of that, the term Little Green Men became synonymous with extraterrestrial beings.

Nordic Aliens

Nordic aliens are extraterrestrial beings that resemble Nordic-Scandinavians. According to UFO stories, Nordic Aliens have great physique. They are usually two meters tall, and have blue eyes and blond hair. As opposed to the skin colors of other extraterrestrial beings mentioned before, they have tanned or fair colored skin. Alternatively, these beings are also called Pleiadians – beings who originated from the Pleiades system.

Back then, they were called Space Brothers. They were most probably called by that name since most of the Nordic Aliens mentioned in close encounter stories were male. Most of the stories and reports about them are usually from contactees. Together with their pleasant appearance, they are also good-natured, which is in total contrast to the common impression of people towards aliens.

One of the prime reasons Nordic Aliens talk to their contactees is their concern about the planet Earth. They want to help humans in preserving Earth's peace and environment. According to John Carpenter (an American social worker who interviewed Nordic Alien contactees), most of them had a pleasant and positive experience with those extraterrestrial beings. Few of the most common traits that contactees said about Nordic Aliens are all knowing, youthful, affectionate, cheerful, watchful, and paternal. On the other hand, the contactees often see these aliens as their protectors and spiritual guides.

Aside from conversing about the Earth's wellbeing, some have said that Nordic Aliens often warn people about Greys. However, that information contradicts the testimonies of other contactees who have seen Nordic Aliens and Greys on board the same UFO. As mentioned in some accounts, it is possible that Nordic Aliens are the leaders and Greys are their subordinates.

On a different note, some experts conclude that the existence of Nordic Aliens was created to fulfill some people's wish to meet a good willed extraterrestrial being. As one can see, the physical characteristics of Nordic Aliens are as pleasant as their goals and demeanors. On the other hand, the beings that are described as ugly are considered malevolent or harmful to the human race.

At present, reports about these beings are rare. It might have something to do with man's perception on aliens. As mentioned before, many people have the idea that aliens are green, gray, or metallic humanoids. Since Nordic Aliens are completely different from such creatures, it is probable that their existence is unknown to many. Hence, some of the hoaxes do not include them. That situation lessens the chances of more Nordic Alien related reports appearing.

Reptilians

As their name suggests, Reptilians share many characteristics with cold-blooded animals. Fundamentally, they are reptilian humanoids that are usually portrayed in science fiction books and films. They are also called draconians, reptiloids, and reptoids.

The concept of Reptilians was brought forth by a popular conspiracy theorist named David Icke. According to him, these Reptilians have the ability to shape-shift, and their main goal is to gain control over human societies. They usually transform into humans to take an active role in politics for them to achieve their main goal. Being a conspiracy theorist, David Icke believes that many political figureheads nowadays are actually Reptilians or being possessed or controlled by them.

Sirians

According to Robert K. G. Temple's Sirius Mystery, it is possible that the Dogon people who live in Mali, Africa have constant communication with extraterrestrial beings. Those beings are Sirians who are located in the Sirius star system.

The book further elaborates that the Sirians are the ones responsible for teaching humans on how to progress in life. Few of the evidences that the author linked to his hypothesis are the Epic of Gilgamesh, mythologies of Greek civilizations, and the systems that the Pharaohs of Egypt used. To support the claim, Temple compared the similarities of Dogon, Sumerian, and Egyptian symbols and beliefs. He even added the correlation of Arab and Greek myths to his claim.

However, Temple also provided other possibilities or events that might have caused the advancements of the parties involved in theory. Nevertheless, even though he is not 100% in full support of his Sirian theory, he believes that it has the highest chance of being true.

Tall Whites

According to accounts from Charles Hall, Tall Whites are extraterrestrial beings that occupy Area 52 of Nellis Air Force Base, Nevada. He encountered them multiple times in that place during his two-year stay. Unlike humans, Tall Whites are fragile creatures. At first glance, Tall Whites seem incapable of talking; and one might think that they are using telepathy to communicate. However, they can talk, and cannot communicate telepathically. Actually, their vocal range is too high for humans to hear and understand what they are saying. It is possible that dogs can hear the pitch of their voices. Since they talk with a high pitch, it is sensible to assume that they also have sensitive hearing organs.

Even without telepathic powers, they are capable of reading the human mind. For them to do that, they will need to equip themselves with a certain electronic communication device. During his encounters, Hall met a few Tall Whites that can understand English. In the event that a Tall White cannot understand English, it will try to communicate by utilizing hand gestures.

On the other hand, Hall believes that the reason Tall Whites are in Area 52 is that they have an agreement with the government. In exchange for technology from the Tall Whites, the government would provide them with human technology or information they do not have. However, Tall Whites are not willing to give technologies that will allow humans to travel in space or create advance forms of weaponry.

UFO Encounters

Sightings and UFO reports are categorized for ufologists and experts to separate plausible stories from hoaxes. Technically, the two main categories are close encounters and distant encounters. These categories were made by J. Allen Hynek, who based the categorization on the distance of the eyewitness from the UFO. By the way, J. Allen Hynek is an avid UFO researcher and well-known astronomer. He published a book about close encounters in 1972. The book's title is *The UFO Experience: A Scientific Inquiry*.

Anyway, close encounter are sightings within the range of 500 feet between the witness and the UFO. On the other hand, the encounter can be considered flawed if the witness is more than 500 feet away from the UFO. These kinds of encounters are usually doubted since the possibility of misidentification is relatively high with that great distance. Few of the UFOs that are not considered close encounters are usually deliberated as human flying aircrafts or natural phenomena.

More often than not, distant encounters are disregarded immediately. But when it comes to close encounters, experts will immediately check on them, especially if the witness has evidences to prove his story. In this case, Hynek has created a scale that he uses to determine how spectacular an encounter is. Without further ado, here is the Hynek's Scale.

Close Encounters

Close Encounter Scale – Original

First Kind

A close encounter is considered a First Kind encounter if the witness only saw the UFO. These are the most commonly reported UFO encounters, and the ones that receive disdain from people.

Second Kind

An encounter is considered a Second Kind if the witness was able to bring or present physical evidence. This kind of encounters is supported by evidence such as a photo or a video. This type of encounters is usually subjected to scrutiny. However, if the evidences are the likes of unfamiliar mechanical parts, unreasonable vegetation damage, sudden electromagnetic interference, or weird soil depressions, the case becomes instantly hot and controversial.

Third Kind

To be categorized as a Third Kind encounter, the witness must have experienced being inside or relatively near the UFO. It is typical that witnesses who have experienced close encounters of the Third Kind tend to write books about their experiences. Naturally, it is difficult to disprove their claims; and on most occasions, they are subjected to interrogations or faced with many people who want to validate their experiences. Also, they are often labeled as mentally ill or fame seeking people.

Close Encounter Scale – Extended

Fourth Kind – addition made by Haynek

Haynek added the Fourth Kind later. To be considered as a Fourth Kind encounter, the witness should experience human abduction. While it might sound unreal, it has been suggested that it is possible that five to six percent of the general population have already experienced alien abduction. In contrast to that, only 1,700 people have claimed that aliens have abducted them. Unfortunately, most of the claimants provide similar abduction narrative patterns, which make them less reliable and believable.

Fifth Kind – addition made by the scientific community

Most popular cases are close encounters of the Fifth Kind. A case can be considered of the Fifth Kind in case the witness was able to communicate with the extraterrestrial being; and in the event the witness was able to maintain contact for long periods, they can become famous overnight. Most of the convincing and detailed stories came from people who experienced a Fifth Kind close encounter.

Distant Encounters

Distant Encounters can be classified into three types. Of course, all of these types are lesser in magnitude than the First Kind close encounters, and they are mainly sightings of UFOs.

Nocturnal Lights

UFO seen during the night, it is usually described as anomalous or weird lights present in the sky during the sighting. Most of the time, they are proven to have been generated by manmade lights.

Daylight Discs

These are sightings during the day. The objects sighted are usually unfamiliar and weird. Also, the flying object does not need to be like a flying saucer to be categorized in this distant encounter type.

Radar/Visual Cases

Under this type are events where a witness sees the anomalous flying object, and at the same time, radar has picked it up. This is far more credible than the first two; and it is usually more believable that a First Kind encounter.

Conclusion

Thank you again for downloading this book!

I hope this book was able to help you to realize who might be at the helm of those UFOs.

Here is some additional food for thought. There are six possibilities on what or who could be the pilots of UFOs.

The UFO is actually a manmade object that a human pilots.

The UFO is actually a manmade object that is unmanned.

The UFO is actually an extraterrestrial object that a human pilots.

The UFO is actually an extraterrestrial object that an alien pilots.

The UFO is actually an extraterrestrial object that runs without pilots.

The UFO is actually an alien.

With the information you have gained today, you will be able to come up with your own conclusion; and with those six possibilities, which one are you leaning towards? Which do you think is the most plausible scenario?

Unfortunately, there is another possibility – a possibility that UFOs do not really exist. You should think why UFOs sightings suddenly became prevalent when humanity is already on the verge of the discovery of highly advanced technologies. Why do UFO talks suddenly emerge during wars? And why have these same talks been dead for some time now? It is up to you to find the answers to those questions.

www.ingramcontent.com/pod-product-compliance
Lightning Source LLC
Chambersburg PA
CBHW060407190526
45169CB00002B/785